ANIMALS OF THE DESERT

Thorny Devils

by Patrick Perish

BLASTOFF!
2
READERS

BELLWETHER MEDIA · MINNEAPOLIS, MN

Blastoff! Readers are carefully developed by literacy experts to build reading stamina and move students toward fluency by combining standards-based content with developmentally appropriate text.

Level 1 provides the most support through repetition of high-frequency words, light text, predictable sentence patterns, and strong visual support.

Level 2 offers early readers a bit more challenge through varied sentences, increased text load, and text-supportive special features.

Level 3 advances early-fluent readers toward fluency through increased text load, less reliance on photos, advancing concepts, longer sentences, and more complex special features.

★ **Blastoff! Universe**

Reading Level

Grade **K**

Grades **1–3**

Grade **4**

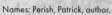

This edition first published in 2021 by Bellwether Media, Inc.

No part of this publication may be reproduced in whole or in part without written permission of the publisher. For information regarding permission, write to Bellwether Media, Inc., Attention: Permissions Department, 6012 Blue Circle Drive, Minnetonka, MN 55343.

Library of Congress Cataloging-in-Publication Data

Names: Perish, Patrick, author.
Title: Thorny devils / by Patrick Perish.
Description: Minneapolis, MN : Bellwether Media, Inc., 2021. | Series:
   Blastoff! readers: animals of the desert | Includes bibliographical
   references and index. | Audience: Ages 5-8 | Audience: Grades K-1 |
   Summary: "Relevant images match informative text in this introduction to
   thorny devils. Intended for students in kindergarten through third
   grade"-- Provided by publisher.
Identifiers: LCCN 2019054273 (print) | LCCN 2019054274 (ebook) | ISBN
   9781644872222 (library binding) | ISBN 9781618919809 (ebook)
Subjects: LCSH: Agamidae--Arid regions--Juvenile literature.
Classification: LCC QL666.L223 P47 2021 (print) | LCC QL666.L223 (ebook) | DDC 597.95/5--dc23
LC record available at https://lccn.loc.gov/2019054273
LC ebook record available at https://lccn.loc.gov/2019054274

Editor: Rebecca Sabelko   Designer: Josh Brink

Printed in the United States of America, North Mankato, MN.

Table of Contents

Life in the Desert   4

A Devil's Tricks   14

Bugs for Breakfast   18

Glossary   22

To Learn More   23

Index   24

Life in the Desert

spike

Thorny devils are **reptiles** with a lot of **spikes**.

They are **adapted** to
the dry deserts of Australia.
This **biome** has few grasses
and shrubs.

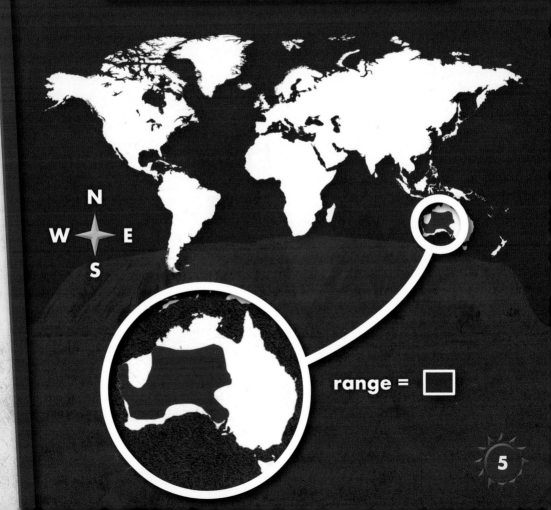

Thorny Devil Range

range =

The desert has little water.
But thorny devils still drink!

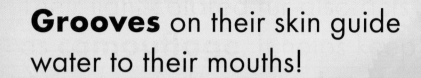

Grooves on their skin guide water to their mouths!

groove

Temperatures change quickly in the desert. Thorny devils' skin color changes, too.

They become a light color in the hot sun. This helps keep them cooler.

Special Adaptations

fake head

spikes

skin grooves

camouflage markings

The desert is filled with hungry **predators**. Thorny devils have a fake head on their backs.

They tuck their real head between their legs to trick enemies.

fake head

These lizards have amazing **camouflage**! Their brown and gold markings match their desert home.

Spikes on their bodies tell
enemies to stay away.

Thorny devils have a special way of walking. They rock back and forth.

This may **confuse** predators.

Thorny Devil Stats

Least Concern	Near Threatened	Vulnerable	Endangered	Critically Endangered	Extinct in the Wild	Extinct

conservation status: least concern

life span: up to 6 years

These lizards dig small
burrows in the sand.

They rest in these burrows when it is too hot or cold.

Bugs for Breakfast

Thorny devils eat a lot of ants!
They use their fast tongues
to gobble up the **insects**.

They chew the ants with special teeth.

Thorny Devil Diet

acrobat ants

carpenter ants

rainbow ants

19

Thorny devils sit along ant trails. They wait for their food to come to them.

These ant eaters make
desert life look easy!

Glossary

adapted—well suited due to changes over a long period of time

biome—a large area with certain plants, animals, and weather

burrows—holes or tunnels some animals dig for homes

camouflage—coloring or markings that make animals look like their surroundings

confuse—to make something difficult to understand

grooves—long, thin dips in the skin of thorny devils

insects—small animals with six legs and hard outer bodies; an insect's body is divided into three parts.

predators—animals that hunt other animals for food

reptiles—cold-blooded animals that have backbones and lay eggs

spikes—pointed parts of the skin of thorny devils

temperatures—measurements of hot and cold

To Learn More

AT THE LIBRARY

Fletcher, Patricia. *Why Do Thorny Devils Have Two Heads?* New York, N.Y.: Gareth Stevens Publishing, 2017.

Perish, Patrick. *Collared Lizards*. Minneapolis, Minn.: Bellwether Media, 2019.

Zappa, Marcia. *Thorny Devils*. Minneapolis, Minn.: ABDO Publishing, 2016.

ON THE WEB

FACTSURFER

Factsurfer.com gives you a safe, fun way to find more information.

1. Go to www.factsurfer.com.

2. Enter "thorny devils" into the search box and click 🔍.

3. Select your book cover to see a list of related content.

Index

adaptations, 5, 9

Australia, 5

biome, 5

bodies, 13

burrows, 16, 17

camouflage, 9, 12

color, 8, 9, 12

dig, 16

food, 18, 19, 20

grooves, 7, 9

head, 9, 10, 11

insects, 18

legs, 10

lizards, 12, 16

markings, 9, 12

mouths, 7

predators, 10, 14

range, 5

reptiles, 4

skin, 7, 8, 9

spikes, 4, 9, 13

status, 15

sun, 9

teeth, 19

temperatures, 8

tongues, 18

walking, 14

water, 6, 7

The images in this book are reproduced through the courtesy of: Stu's Images, front cover (hero); bmphotographer, front cover (background), pp. 2-3; EyeEm/ Alamy, p. 4; Natural Visions/ Alamy, p. 6; Uwe Bergwitz, pp. 7, 13; crbellette, p. 8, 22; Morales/ Alamy, p. 9; Gone For A Drive, p. 9 (top); Alessandra Sarti/ SuperStock, p. 10; Bildagentur Zoonar GmbH, p. 11; witte-art_de, p. 12; Jonathan Ayres/ Alamy, p. 14; Marcelo_Photo, p. 15; Layer, W./ Newscom, p. 16; Martin Harvey/ Newscom, p. 17; Fritz Hiersche, p. 18; Young Swee Ming, p. 19 (top right); Peter Yeeles, p. 19 (top left); Steve Shattuck, p. 19 (bottom); Leith Holtzman, p. 20; Mitsuaki Iwago/ SuperStock, p. 21.